Contents

Chapter 1:
The Imperative
for Change

Arshia Shayestehtabar

Sustainable
Future

Creating a Better World
for Future Generations

The Imperative for Change

In a world facing unprecedented environmental challenges, the imperative for change has never been clearer or more pressing. Our planet, the one and only home we have, is in peril. It is a call to action that resonates across borders, cultures, and generations. The urgency of this call emanates from a deep understanding of the environmental challenges we face, the consequences of inaction, and the critical role that sustainability plays in securing a better future for all.

The challenges before us are immense and multifaceted. We find ourselves at a crossroads, where the decisions we make today will shape the world we leave for future generations. The first step in addressing these challenges is recognizing the pressing need for change. It requires acknowledging that our current path is unsustainable and that the status quo is no longer an option.

One of the most critical aspects of the imperative for change is the recognition of the environmental challenges we face. Our planet is experiencing disruptions in its natural systems at an alarming rate. Climate change, driven by the excessive emission of greenhouse gases, is altering our weather patterns, causing more frequent and severe natural disasters, and threatening the stability of ecosystems. Rising sea levels and extreme weather events are already affecting communities around the world, displacing populations and wreaking havoc on economies.

Biodiversity loss is another grave concern. The rapid decline in the number of species on Earth is not just a loss of beauty and wonder; it disrupts the delicate balance of ecosystems and can lead to unforeseen consequences for human life. The loss of pollinators, for example, threatens our food supply, as many of our staple crops depend on insect pollination.

The imperative for change also encompasses the degradation of our natural resources. Our relentless exploitation of forests, fisheries, and freshwater sources is depleting these vital resources faster than they can regenerate. This not only threatens the livelihoods of countless people who depend on these resources but also undermines the very foundation of life on Earth.

As we confront these challenges, we must also grapple with the consequences of inaction. The cost of doing nothing is steep. Continued environmental degradation will lead to more frequent and severe crises, both in terms of human suffering and economic losses. The burden will fall disproportionately on the most vulnerable among us, exacerbating inequalities and social unrest.

Furthermore, failing to address these challenges now will limit our options in the future. It will narrow the window of opportunity for effective solutions and force us into a perpetual state of crisis management. The longer we delay action, the more drastic and disruptive the changes required will become.

It is against this backdrop that the role of sustainability becomes paramount. Sustainability is not merely a buzzword or a trendy concept; it is a fundamental principle that must guide our actions. At its core, sustainability is about meeting the needs of the present without compromising the ability of future generations to meet their needs. It is about striking a balance between economic development, social equity, and environmental stewardship.

Sustainability offers a path forward, a way to address the imperative for change comprehensively and effectively. It provides a framework for rethinking our relationship with the planet and reimagining the way we live, work, and interact with one another. It challenges us to question the status quo and to seek innovative solutions that benefit not only our generation but also those that will come after us.

Embracing sustainability means making choices that reduce our ecological footprint. It means transitioning to clean and renewable sources of energy, adopting circular economies that minimize waste, and reimagining transportation systems that are both efficient and environmentally friendly. It means valuing and protecting our natural resources, recognizing the intrinsic value of biodiversity, and conserving the beauty and wonder of our planet.

Sustainability is not an one-size-fits-all solution; it is a guiding principle that can be applied across industries, sectors, and communities.

Whether you are in business, government, academia, or a concerned citizen, there is a role for you to play in advancing sustainability. It is a collective effort that requires collaboration, innovation, and a shared commitment to change.

Chapter 2:
The Foundations
of Sustainability

The Foundations of Sustainability

Sustainability is not a fleeting trend or a passing buzzword; it is a fundamental concept that shapes the way we interact with the world around us. To understand sustainability fully, we must explore its foundations definition, historical perspectives, and the triple bottom line approach.

Defining Sustainability

At its core, sustainability is about balancing the needs of the present with the capacity of the planet to support those needs while ensuring that future generations can do the same. It is a recognition that our actions have consequences, and that we have a responsibility to consider the long-term impacts of our choices. Sustainability challenges us to find harmony between environmental health, social equity, and economic prosperity.

Environmental sustainability involves minimizing our negative impacts on the natural world and preserving the Earth's ecosystems. It means reducing pollution, conserving resources, and mitigating climate change. It's about safeguarding biodiversity and respecting the intricate web of life that sustains us all.

Social sustainability focuses on people and communities. It means ensuring that everyone has access to the resources they need to live a dignified life. This includes access to clean water, nutritious food, education, healthcare, and more. Social sustainability also entails promoting inclusivity, equity, and justice, so that no one is left behind.

Economic sustainability seeks to create prosperity without compromising the well-being of future generations. It involves responsible resource management, fair labor practices, and a commitment to long-term economic stability. It encourages innovation and efficiency while considering the social and environmental costs of economic activities.

Historical Perspectives

The concept of sustainability is not a recent invention. Indigenous cultures around the world have practiced sustainable living for centuries. They understood the delicate balance between human needs and the Earth's capacity to provide. Their practices often revolved around respecting nature, conserving resources, and passing down knowledge through generations.

In the modern context, the term «sustainability» gained prominence in the 20th century as environmental concerns grew. The 1960s and 1970s saw the birth of the environmental movement, with events like Earth Day highlighting the need to protect our planet. The 1980s brought discussions about sustainable development, linking environmental and social concerns to economic growth.

In 1987, the Brundtland Report, commissioned by the United Nations, introduced the world to the concept of sustainable development. It defined sustainable development as «development that meets the needs of the present without compromising the ability of future generations to meet their own needs.» This report marked a turning point in global discussions about sustainability, emphasizing the interdependence of economic, social, and environmental factors.

The Triple Bottom Line Approach

To operationalize sustainability, organizations and businesses have adopted the triple bottom line approach. This framework evaluates success not just in terms of profit but also in terms of social and environmental impact. It recognizes that a narrow focus on financial gain can come at the expense of other crucial factors.

The three «bottom lines» of the triple bottom line approach are:

Profit (Economic Bottom Line): This measures the financial success of an organization. It involves considerations of revenue, profit margins, and return on investment. However, it also considers whether profit is being generated through ethical and sustainable practices.

People (Social Bottom Line): This focuses on the impact of an organization on society. It considers employee well-being, community engagement, and social responsibility. A socially responsible organization seeks to make a positive contribution to the communities it operates in.

Planet (Environmental Bottom Line): This assesses the environmental impact of an organization's activities. It looks at resource use, emissions, waste generation, and environmental conservation efforts. An environmentally responsible organization strives to minimize its ecological footprint.

The triple bottom line approach encourages organizations to find synergy between these three dimensions. It acknowledges that financial success should not come at the expense of people or the planet. Instead, it promotes a holistic view of sustainability, where economic prosperity supports social well-being and environmental stewardship.

In this chapter, we've explored the foundations of sustainability, the definition that encompasses environmental, social, and economic considerations, its historical roots, and the triple bottom line approach that operationalizes it. These foundations provide a solid framework for understanding how sustainability can guide our actions and decisions in various domains, from agriculture and energy to business and policymaking.

As we move forward, we will delve deeper into practical aspects of sustainability, examining sustainable practices across different industries and sectors. We will explore how individuals, organizations, and governments can implement sustainable solutions that not only benefit the present but also secure a better future for generations to come. The journey to building a sustainable future is grounded in these foundational principles, serving as our guide in the chapters ahead.

Chapter 3: Sustainable Practices

Sustainable Practices

To create a better world for future generations, we must translate the principles of sustainability into practical, everyday practices. In this chapter, we will explore sustainable practices across various domains, from agriculture and energy to waste reduction and transportation.

Sustainable Agriculture

Agriculture is at the heart of our food system, and adopting sustainable practices in this sector is essential for both food security and environmental health. Sustainable agriculture seeks to minimize the environmental impact of farming while ensuring a resilient and productive food supply.

One key approach is organic farming, which avoids synthetic pesticides and fertilizers, opting instead for natural alternatives. Organic farming promotes soil health, reduces chemical runoff, and protects biodiversity. It also encourages crop rotation and diversification, which enhance the resilience of farming systems.

Another sustainable practice is precision agriculture, which leverages technology like GPS and sensors to optimize resource use. By precisely monitoring and managing factors like water, fertilizer, and pesticides, farmers can reduce waste and minimize the environmental footprint of their operations.

Renewable Energy Sources

Transitioning to renewable energy sources is a cornerstone of sustainability in the energy sector. Fossil fuels, such as coal and oil, are finite resources that contribute to air pollution and climate change. In contrast, renewable energy sources like solar, wind, and hydropower harness the Earth's natural processes without depleting finite reserves or emitting harmful greenhouse gases.

Solar energy, for example, captures the power of sunlight through photovoltaic panels, converting it into electricity. Wind turbines generate electricity from the kinetic energy of wind. Hydropower systems harness the energy of flowing water, such as rivers and streams. These renewable sources offer a cleaner, more sustainable alternative to fossil fuels.

Waste Reduction and Recycling

Waste reduction and recycling are essential components of sustainable resource management. They reduce the burden on landfills and conserve valuable materials.

Waste reduction starts with source reduction, which involves minimizing waste at its source. This can include using products with less packaging, reusing items, and practicing responsible consumption. By producing and purchasing less, we reduce the waste generated in the first place.

Recycling is another crucial practice. It involves collecting and processing materials like paper, glass, plastics, and metals to be used as raw materials for new products. Recycling conserves energy and resources compared to producing items from virgin materials.

Eco-Friendly Transportation

The transportation sector is a significant contributor to greenhouse gas emissions and air pollution. Adopting eco-friendly transportation practices is vital to reduce the sector 's impact on the environment.

One approach is promoting public transportation and producing individual car use. Efficient and accessible public transit systems reduce traffic congestion and lower emissions per passenger-mile traveled. Carpooling and sharing also help optimize vehicle occupancy.

Transitioning to electric vehicles (EVs) is another sustainable transportation practice. EVs produce zero tailpipe emissions and are more energy-efficient than traditional gasoline-powered vehicles. Additionally, sustainable urban planning can lead to more walkable and bike-friendly cities, reducing the reliance on cars.

Conservation of Natural Resources

Conservation of natural resources is a fundamental sustainable practice that involves responsible stewardship of our planet 's finite resources. It encompasses practices like sustainable forestry, fisheries management, and water conservation.

Sustainable forestry focuses on harvesting trees in a way that maintains the health and diversity of forest ecosystems. It involves selective logging, reforestation, and protection of old-growth forests.

Fisheries management aims to prevent overfishing and preserve marine ecosystems. Sustainable practices include setting catch limits, enforcing fishing regulations, and protecting critical habitats.

Water conservation practices, such as efficient irrigation systems and reduced water waste, are essential to ensure access to clean water for future generations. Responsible water management helps mitigate the impact of water scarcity and drought.

Environmental Stewardship in Business

Incorporating sustainability into business operations is not only an ethical choice but also a strategic one. Companies that adopt sustainable practices often benefit from reduced operational costs, increased market share, and improved brand reputation.

Business sustainability involves reducing energy consumption, minimizing waste, and adopting eco-friendly production processes. Companies can also engage in sustainable sourcing, ensuring that the materials and resources they use are obtained responsibly.

Additionally, transparency and accountability play a critical role in business sustainability. Reporting on environmental and social impacts, as well as setting sustainability goals, helps companies track their progress and communicate their commitment to stakeholders.

In this chapter, we 've explored a range of sustainable practices that can be applied in various sectors of our lives, from agriculture and energy to waste reduction and transportation. These practices are not isolated efforts; they form a web of interconnected solutions that contribute to building a more sustainable future. As we continue our journey, we will delve deeper into the concept of corporate responsibility in the context of sustainability, understanding how businesses can play a pivotal role in shaping a better world for future generations.

Chapter 4: Corporate Responsibility

Corporate Responsibility

Corporate responsibility is more than a buzzword; it's a fundamental principle that guides the actions of businesses in a rapidly changing world. In this chapter, we will explore the concept of corporate responsibility in the context of sustainability, examining how organizations can play a pivotal role in creating a better world for future generations.

The Business Case for Sustainability

Sustainability and corporate responsibility are not just about doing good; they also make good business sense. Forward-thinking companies recognize that aligning their operations with sustainability principles can lead to long-term success and resilience.

Risk Mitigation: Embracing sustainability practices helps companies mitigate risks associated with environmental and social issues. For example, reducing greenhouse gas emissions can lower exposure to climate-related risks, such as extreme weather events and supply chain disruptions.

Cost Savings: Sustainability often goes hand in hand with cost savings. Improving energy efficiency, reducing waste, and conserving resources can lead to significant operational efficiencies and lower expenses.

Market Opportunity: Consumers increasingly seek out sustainable products and services. Companies that prioritize sustainability can tap into growing market segments and gain a competitive edge.

Brand Reputation: A commitment to corporate responsibility enhances brand reputation. It demonstrates that a company values ethical business practices and is willing to address environmental and social challenges.

Environmental Responsibility

Environmental responsibility is a key pillar of corporate sustainability. Companies can take several actions to minimize their environmental impact:

Reducing Emissions: Many organizations are setting ambitious goals to reduce their carbon emissions. This involves transitioning to renewable energy sources, optimizing transportation logistics, and implementing energy-efficient technologies.

Resource Efficiency: Businesses can adopt circular economy principles to reduce waste and extend the life cycle of products. This includes designing products for easy disassembly and recycling.

Sustainable Sourcing: Companies can ensure that the materials and resources they use are obtained responsibly. Sustainable sourcing practices can help protect ecosystems and support communities.

Social Responsibility

Social responsibility encompasses a wide range of practices that promote the well-being of employees, communities, and society at large:

Fair Labor Practices: Treating employees with fairness and respect is a fundamental aspect of social responsibility. This includes paying fair wages, ensuring safe working conditions, and supporting work-life balance.

Community Engagement: Engaging with local communities helps build trust and foster positive relationships. Companies can support local initiatives, contribute to education, and provide disaster relief.

Diversity and Inclusion: Embracing diversity and inclusion within the workplace is not only a moral imperative but also a source of innovation and competitiveness. Companies can create inclusive environments where all employees thrive.

Transparency and Accountability

Transparency and accountability are essential components of corporate responsibility. Companies that are transparent about their sustainability efforts are more likely to build trust with stakeholders.

Reporting: Many organizations publish sustainability reports that detail their environmental and social performance. These reports provide a transparent view of a company's actions and progress.

Setting Goals: Setting specific sustainability goals and targets help organizations stay accountable. These goals can relate to emissions reductions, waste reduction, diversity initiatives, and more.

Stakeholder Engagement: Engaging with stakeholders, including employees, customers, investors, and communities, is a crucial part of corporate responsibility. Companies can seek input, listen to concerns, and incorporate feedback into their strategies.

Sustainability in the Supply Chain

The supply chain is a critical area where companies can exert significant influence on sustainability. Many businesses are taking steps to ensure that their supply chains align with sustainability principles.

Supplier Audits: Conducting audits of suppliers helps ensure that they meet environmental and social standards. This practice helps identify areas for improvement and promotes responsible sourcing.

Supply Chain Transparency: Transparency in the supply chain involves tracing the origin of materials and ensuring that they meet ethical and environmental standards. This transparency can help identify and address issues like deforestation and human rights violations.

Sustainable Procurement: Companies can prioritize procurement from suppliers who adhere to sustainable practices. This encourages suppliers to adopt more responsible behaviors.

Sustainable Innovation

Innovation plays a vital role in corporate responsibility. Companies are harnessing their resources and expertise to develop innovative solutions that address pressing global challenges:

Product Innovation: Creating products and services with sustainability in mind can lead to breakthroughs that benefit society and the environment. This includes developing eco-friendly materials, energy-efficient technologies, and renewable energy solutions.

Collaborative Innovation: Collaboration across industries, sectors, and even competitors can drive sustainable innovation. Partnerships can lead to the development of transformative technologies and solutions.

Circular Business Models: Some organizations are shifting from linear business models (produce, use, dispose) to circular models (produce, reuse, recycle). This reduces waste and promotes resource efficiency.

The Role of Leadership

Leadership within organizations is critical for driving corporate responsibility. Leaders set the tone, define the company's values, and inspire employees to embrace sustainability:

Visionary Leadership: Leaders who articulate a compelling vision for sustainability inspire employees to get involved and work toward common goals.

Lead by Example: Leading by example means incorporating sustainability practices into the company 's culture and operations. It involves making sustainable choices at all levels of the organization.

Empower Employees: Encouraging employees to contribute to sustainability efforts fosters a sense of ownership and engagement. Employees often have valuable insights and ideas for sustainability improvements.

In this chapter, we've explored the concept of corporate responsibility within the context of sustainability. We 've seen how organizations can play a pivotal role in creating a better world for future generations by embracing environmental and social responsibility, promoting transparency and accountability, and driving sustainable innovation. The journey to a sustainable future is a collective effort and businesses are key players in this transformative process. As we continue, we 'll delve into the importance of inclusivity in sustainable initiatives and explore how governments, NGOs, and individuals can contribute to building a better world.

Chapter 5: Fostering Inclusivity

Fostering Inclusivity

Inclusivity is a fundamental aspect of sustainability and corporate responsibility. It goes beyond environmental and economic considerations, encompassing the social dimension of sustainability. In this chapter, we will explore the importance of fostering inclusivity in sustainable initiatives and how it contributes to creating a better world for future generations.

The Inclusive Approach to Sustainability

Sustainability is not only about protecting the environment and managing resources wisely; it 's also about ensuring that the benefits of sustainable practices reach all members of society. An inclusive approach recognizes that different communities and individuals have unique needs, challenges, and perspectives. It seeks to address disparities and promote equity in all aspects of sustainability.

Social Equity and Inclusion

Promoting social equity and inclusion is a central aspect of fostering inclusivity in sustainability. Here are key considerations:

Access to Resources: Inclusive sustainability ensures that all individuals and communities have equitable access to resources essential for a dignified life. This includes clean water, nutritious food, education, healthcare, and affordable housing.

Empowerment: Inclusivity empowers marginalized communities by involving them in decision-making processes. Their insights and experiences are invaluable for crafting sustainable solutions that address their unique challenges.

Diverse Perspectives: Inclusive sustainability values and incorporates diverse perspectives. This diversity of thought often leads to innovative solutions and more comprehensive approaches to sustainability challenges.

Gender Equality in Sustainability

Gender equality is a crucial dimension of inclusivity in sustainability. Women play pivotal roles in many communities, and empowering them is essential for achieving sustainable development. Consider the following:

Economic Empowerment: Sustainable initiatives should prioritize economic opportunities for women, including access to sustainable livelihoods, fair wages, and entrepreneurial opportunities.

Education and Health: Inclusive sustainability ensures that women have access to quality education and healthcare. Investing in women 's health and education benefits families and communities.

Leadership Roles: Encouraging women to take on leadership roles in sustainability initiatives and organizations promotes gender equity and enriches decision-making processes.

Community Engagement

Inclusivity also involves engaging with local communities in sustainability initiatives. Here's how this can be achieved:

Participatory Planning: Involving local communities in the planning and decision-making processes of sustainability projects ensures that initiatives align with their needs and priorities.

Capacity Building: Empowering communities through training and skill development helps them take an active role in implementing and maintaining sustainability projects.

Benefits Sharing: Inclusivity means that communities benefit from the positive outcomes of sustainability initiatives, such as job creation, improved living conditions, and increased access to resources.

Inclusivity in Policy and Governance

Governments and policymakers play a crucial role in fostering inclusivity in sustainability. Some key considerations include:

Equitable Policies: Policies should be designed with an equity lens to ensure that they do not inadvertently harm marginalized communities and that they promote social inclusion.

Representation: Governments should ensure that diverse voices are represented in decision-making bodies, reflecting the demographics of the population.

Access to Justice: Legal frameworks should provide marginalized communities with access to justice, enabling them to protect their rights and interests.

Indigenous Knowledge and Traditional Practices

Indigenous communities often hold valuable knowledge about sustainable practices and environmental stewardship. Fostering inclusivity in sustainability means recognizing and respecting this traditional knowledge. Indigenous communities involvement can lead to more holistic and culturally sensitive approaches to sustainability.

Inclusivity in Education and Awareness

Education and awareness campaigns about sustainability should be inclusive and accessible to all. This includes:

Accessible Information: Providing information in multiple languages and formats to ensure that people of different backgrounds can access and understand sustainability concepts.

Engaging Communities: Engaging with communities through workshops, outreach programs, and educational initiatives to build knowledge and knowledge about sustainability.

Youth Engagement: Inclusivity in sustainability involves empowering young people to take an active role in advocating for and implementing sustainable practices.

Measuring Inclusivity

Measuring inclusivity in sustainability can be challenging but is essential for tracking progress. Metrics should go beyond traditional economic indicators and encompass social and environmental dimensions. Key performance indicators may include measures of social equity, access to resources, and the participation of underrepresented groups in decision-making processes.

Challenges and Barriers to Inclusivity

Fostering inclusivity in sustainability initiatives is not without challenges. These may include:

Power Imbalances: Overcoming power imbalances and addressing historical injustices often requires significant effort and collaboration.

Cultural Sensitivity: Ensuring that sustainability initiatives are culturally sensitive and respect the traditions and values of different communities can be complex.

Resource Constraints: Limited resources may pose challenges to inclusivity efforts, especially in resource-constrained environments.

The Path Forward

Fostering inclusivity in sustainability is not only a moral imperative but also a strategic choice. It enhances the effectiveness and impact of sustainability initiatives, ensures that the benefits of sustainability reach all members of society, and contributes to a more equitable and just world for future generations.

As we continue our journey toward building a sustainable future, we must prioritize inclusivity in all aspects of sustainability, from policy and government to business practices and community engagement. By doing so, we can create a world where everyone has the opportunity to thrive, and no one is left behind.

Chapter 6: Building a Sustainable Future

Building a Sustainable Future

In the preceding chapters, we've explored the imperative for change, the foundations of sustainability, sustainable practices, corporate responsibility, and the importance of fostering inclusivity in sustainable initiatives. Now, in this final chapter, we embark on the journey of building a sustainable future, a journey that requires collective action, innovation, and a steadfast commitment to creating a better world for future generations.

The Collective Effort

Building a sustainable future is a shared responsibility that transcends borders, industries, and individual actions. It requires governments, businesses, communities, and individuals to work together in a collaborative effort. No single entity or solution can address the complex web of sustainability challenges we face.

Policy and Governance

Effective governance and policies at the local, national, and international levels play a pivotal role in driving sustainability. Policymakers have the power to enact laws and regulations that promote sustainable practices, protect the environment, and ensure social equity.

Environmental Protection: Governments can implement and enforce environmental regulations that limit pollution, promote conservation, and combat climate change. International agreements, such as the Paris

Agreement, provide a framework for global cooperation on climate action.

Incentives for Sustainability: Tax incentives, subsidies, and grants can encourage businesses and individuals to adopt sustainable practices. These financial incentives can stimulate investments in renewable energy, energy efficiency, and other sustainable technologies.

Education and Awareness: Governments can support education and awareness campaigns to inform citizens about the importance of sustainability and provide guidance on sustainable behaviors.

Corporate Responsibility

Businesses have a significant role to play in shaping a sustainable future. Besides pursuing profits, companies can prioritize sustainability and social responsibility in their operations. Key areas of focus include:

Supply Chain Responsibility: Businesses can ensure that their supply chains adhere to ethical and environmental standards. This includes sourcing materials responsibly and supporting fair labor practices.

Innovation for Sustainability: Companies can drive innovation by developing sustainable products, services, and technologies. Sustainable innovation often leads to competitive advantages and market differentiation.

Transparency and Reporting: Transparency about sustainability efforts and reporting on environmental and social performance are essential for accountability.

Community Engagement

Engaging with local communities is vital for the success of sustainability initiatives. Communities are often the most directly affected by environmental and social changes. Effective engagement involves:

Participatory Decision-Making: Including community members in decision-making processes ensures that projects align with local needs and priorities.

Capacity Building: Empowering communities through training and skill development enables them to take an active role in sustainability initiatives and project maintenance.

Benefits Sharing: Sustainability initiatives should benefit local communities, creating job opportunities, improving living conditions, and increasing access to resources.

Innovation and Technology

Innovation and technology are powerful drivers of sustainability. Advances in science and technology offer new opportunities for addressing environmental and social challenges.

Clean Energy: Investments in renewable energy sources, such as solar, wind, and hydropower, can transform energy systems and reduce greenhouse gas emissions.

Efficient Transportation: Electric vehicles, public transportation, and sustainable urban planning can reduce the environmental impact of transportation.

Circular Economy: Transitioning to a circular economy, where products are designed for reuse and recycling, can reduce waste and conserve resources.

Individual Actions

Individual actions, while seemingly small in isolation, collectively have a significant impact on sustainability. Each person can contribute to building a sustainable future by:

Conservation: Reducing personal water and energy consumption, conserving resources, and minimizing waste can lower environmental impact.

Responsible Consumption: Making sustainable choices as consumers, such as supporting eco-friendly products and local businesses, can drive market change.

Advocacy: Advocating for sustainability at the community and political levels can influence policies and encourage collective action.

Chapter 7: Conclusion

Conclusion

In conclusion, the journey to a sustainable future is a collective and ongoing effort that requires the commitment and cooperation of individuals, businesses, communities, and governments worldwide. Sustainability is not just a concept; it's a way of thinking and acting that seeks to balance environmental health, social equity, and economic prosperity.

Throughout this book, we have explored the essential components of sustainability, from its foundational principles to practical sustainable practices. We've delved into the role of corporate responsibility in shaping a better world, emphasized the importance of fostering inclusivity in all sustainability efforts, and discussed the critical need for policies, innovations, and individual actions that support sustainability.

Building a sustainable future is not without its challenges and complexities, but it is a journey worth embarking on. The consequences of inaction are too great, affecting not only the present but also the well-being and opportunities of future generations. Climate change, resource depletion, social inequalities, and environmental degradation are pressing issues that demand our attention and commitment to change.

In this journey, we must embrace inclusivity, recognizing that every individual and community has a role to play and deserves the benefits of sustainability. We must continue to innovate, seeking new solutions and technologies that reduce our environmental impact and improve the quality of life for all.

Corporate responsibility should be the norm rather than the exception, with businesses leading the way in ethical and sustainable practices. Governments must enact policies that support and incentivize sustainability, fostering an environment where sustainable choices are accessible and affordable.

As individuals, our choices and actions matter. Each sustainable decision, no matter how small, contributes to the larger goal of a sustainable future. Whether it's reducing our carbon footprint, supporting local and sustainable products, advocating for change, or simply raising awareness, our efforts collectively make a difference.

Education and lifelong learning are the keys to equipping ourselves with the knowledge and skills needed to address the complex challenges of sustainability. By continuously seeking to understand and engage with sustainability concepts, we empower ourselves to make informed decisions and inspire change in our communities.

Ultimately, building a sustainable future is a choice one that requires dedication, perseverance, and a shared vision of a world where people and the planet thrive together. It is a journey that we must undertake together, leaving a legacy of stewardship, responsibility, and compassion for future generations.

In the pages of this book, we've explored the principles and practices of sustainability, but the true journey has just begun. The path to a sustainable future is waiting for us to take the first step, and with each step, we move closer to realizing a world that is healthier, more equitable, and more prosperous for all. The responsibility is ours, and the opportunity is now.

NOTE:

NOTE:

NOTE:

NOTE:

NOTE:

NOTE:

NOTE:

NOTE:

NOTE:

NOTE:

NOTE:

NOTE:

NOTE:

NOTE:

NOTE:

NOTE:

NOTE:

NOTE:

NOTE:

NOTE:

NOTE:

NOTE:

NOTE:

NOTE:

NOTE:

NOTE:

NOTE:

NOTE:

NOTE:

NOTE:

NOTE:

NOTE:

Book designer: Arshia Shayestehtabar

First Edition: 2023

For permission requests, please contact with the writer Throught this email:

arshiashayestehtabar@gmail.com

www.ingramcontent.com/pod-product-compliance
Lightning Source LLC
Chambersburg PA
CBHW062242290526
45794CB00006B/2377